Potted History

The approaching roar of an unseen airliner climbing out of Geneva Airport stopped us both in our tracks! This was the early 'fifties, and only a Constellation or a Douglas DC-6 could be making a racket like that. "It's a DC-6", I asserted, trying to muster all the authority my 11 years would allow, plus a bit more! "Never!" said my older and wiser pal, "It's a Constellation". Much mortified, I refused to accept defeat and said "Well, it sounds like a DC-6, anyway", hoping the heavy machine would stay unseen and leave my assertion untested! But maddeningly, the Connie hove into view above the buildings, triggering a gleeful stream of words from my pal, who ended up with a mocking "You see, I know my planes!" But he did me a favour, because I never again confused the sounds of Connies and DC-6s, and, in time, learned to recognise numerous aircraft by their engine notes. Thank you, Alain Morvan, and, incidentally, this was the only time I was ever less than pleased to see a Connie!

Lockheed's Constellation had a certain magic and mystique about it which probably no airliner has had since, except the early Comet jetliner. The travelling public sensed the Connie's aura, aided and abetted by airline publicity departments, and the feeling filtered down to airminded schoolboys. "It's a Constellation" meant something rather special, and not for nothing did it seem as though there were two kinds of airline, those with Connies and those without! But why was the Connie so special? Because, quite simply, it was way in advance of anything else when it started airline service just after WW2. Its 300 mph cruise made it 70 mph faster than its nearest "competitor", and its 2,200 hp 18-cylinder engines seemed massively powerful when 1,450 hp and 14 cylinders seemed reasonable. Also, the pressurised Connie established itself as a long-distance machine *par excellence*, and by mid-May 1947 over 80 Atlantic return crossings each week were being safely and reliably made by Connies. Not only this, because the Constellation was one of the most handsome airliners ever built. It was the "glamour girl" of the skies. For a time the aura was fully earned, but even when the Douglas DC-6B knocked the Connie off its throne, the aura still persisted. And while the DC-6B's technical excellence and sales success earned it enormous respect, it never seemed to have the Connie's *cachet* or inspire the same reverence — the latter emotion was for Connies only, from airminded young and old alike!

During the 'thirties Lockheed produced some high-performance all-metal monoplane airliners, the Models 10, 12, 14 and 18, with refinements such as retractable undercarriages, variable-pitch propellers and flaps. The first of these was the Model 10 which appeared in 1934, selling extremely well and putting Lockheed in the same advanced technology league as Douglas. Indeed, the Americans were very advanced in civil aviation during the 'thirties, so much so that the Europeans ended up buying aircraft from the New World, because the Old World produced nothing comparable with the Douglas DC-2 and DC-3 or the magnificent Lockheed twins. Even the self-sufficient British succumbed to American airliners when British Airways bought Lockheed 10s and 14s. In 1936, Lockheed brought out the Model 12, smaller than the Model 10 but with the same engines and hence a more sprightly performance, with cruising and maximum speeds rising from 190 mph and 202 mph, respectively, to 212 mph and 225 mph. The Lockheed 12 sold mostly to private and corporate owners because it was too small for the expanding airline market, as was the Model 10 eventually, and Lockheed introduced the larger Model 14 in 1937 as a DC-3 competitor. In the event the Model 14 failed in this respect due to the DC-3's better economics, but the '14 found its market niche and sold well enough. However, the Douglas and Lockheed machines were all twin-engined, and, with the airlines demanding more capacity, the aircraft industry turned its attention to four-engined airliners resulting in such machines as the Boeing Stratoliner and Douglas DC-4.

Lockheed started d[...]
engined airliner, know[...]
nine-cylinder Wright[...]
Whitney R-1830s. Bo[...]
and gave over 1,000 h[...]
powerful for those days. Yet engines were under develop[...]t which would soon make 1,000 hp look relatively puny, and one such engine was the Wright R-3350 two-row radial which first ran in May 1937, giving a massive 2,000 hp or so from its 54.56 litres and 18 cylinders. The original Excalibur airliner had insufficient passenger capacity, range and performance for TWA and Pan American, and Lockheed proposed a larger aircraft powered by four Wright R-3350 engines. Design-work on the new machine started during the summer of 1938, and the Constellation, as it became known, could carry 44 passengers and had an estimated top speed of 360 mph, making it as fast as the very best interceptor fighters of the time! Perhaps this had a parallel in Concorde about 30 years later. Tricycle undercarriages, which we take for granted now, were just beginning to appear in the late 'thirties and the Connie was naturally given this feature. Also, multiple tail-fins were fashionable when the Connie was being designed, thanks, in part, to Lockheed's Model 10, 12 and 14, and the Connie went one better by having three tailfins.

Not surprisingly, the Connie was caught up in WW2 when the USA entered the conflict on 8 December 1941, and plans to use the Connie as an airliner had to be shelved and the machine was ordered as the C-69 military transport for the USAAF. Construction of the prototype had started the previous year, and the C-69 first flew as NX25600 on 9 January 1943 in the hands of Boeing test pilot Edward T. Allen, who had been lent to Lockheed for his experience of flying large four-engined aircraft. Everything about the new C-69 demanded superlatives, and its combination of performance, ultra-modern appearance and the immense power of its four 2,200 hp engines were an order of magnitude different to anything seen before on an airliner. Not only this, because with its span of 123 ft the Connie was an impressively large aircraft when it first appeared. Only 15 C-69s had been delivered by the time WW2 ended, and with the return of peace the C-69 contract was cancelled. Luckily, customers were found for the cancelled aircraft and these were offered as Model 049 airliners, the first customer being TWA which received 27 cancelled C-69s. Ten of them were registered NC86500 to '509, with delivery starting in late 1945, and they were christened *Star of the Mediterranean*, *Star of the Persian Gulf*, *Star of the Pyramids*, *Navajo Sky Chief*, *Star of France*, *Paris Sky Chief*, *Star of Dublin*, *Star of Madrid Sky Chief*, *Star of Athens* and *Star of Africa*. TWA introduced the Connie on 3 December with NC86505 *Paris Sky Chief* making a special flight from New York to Paris, but unfortunately this inaugural machine had a short innings because it crashed on 28 December 1946 making its approach to Shannon Airport, Eire. Another special flight was made, this time by Pan Am, from New York to Bermuda on 14 January 1946. TWA machine NC86511 inaugurated normal New York–Paris flights on 5 February 1946, and that same month Pan Am inaugurated New York–England services with NC88831, but not to Heathrow — instead Pan Am's Connie flew to Hurn Airport, Bournemouth.

Apart from TWA and Pan Am, several airlines bought Connie 049s, among them British Overseas Airways (BOAC) in the absence of anything suitable from the British aircraft industry, and four new machines were delivered during May to July 1946 as G-AHEK to 'N (*Berwick II, Bangor, Balmoral and Baltimore*); these machines were originally intended for the USAAF. A proposal was made to licence-build the Connie in Great Britain by the Bristol company at Filton (now part of British Aerospace), using Bristol Centaurus engines, but this plan fell through. Inevitably, the new Connie suffered from teething troubles, and the type was grounded on 12 July 1946 to eliminate the causes of some in-flight fires, the final one being due to an electrical fault resulting in a fatal crash the previous day; however, the Connie was given a clean bill of health on 23 August and allowed back into service. The Model 049-46 had a take-off weight of 86,250 lb and its 2,200 hp Wright R-3350 engines drove Hamilton Standard three-blade constant-speed propellers. Customers could choose their own seating arrangement, for instance Air France specified 47 passengers and seven crew, while Linea Aeropostal Venezolanos (LAV) went for 51 passengers and five crew. KLM took both 47 and 51 passenger Connies, and American Overseas Airlines opted for 43 passengers and seven crew. However, in time the Connie was given high-density seating for up to 81 passengers, while BOAC reduced capacity to as little as 30 for de luxe transatlantic flights during the mid-fifties.

For a while the Connie was unique in its size class and its nearest "competitor", the unpressurised Douglas DC-4, offered practically no competition at all because, with its modest 1,450 hp Pratt & Whitney R-2000s, it was too small; not only this, because its 230 mph cruise was too low. Long-haul airliner technology had moved on to 2,200 hp engines and 300+ mph cruising speeds, and although there was nothing technically wrong with the DC-4, it was outdated by the Connie in 1945 despite being relatively new. Luckily for Douglas the DC-4 was capable of being enlarged and developed into the pressurised DC-6, with 2,100 hp P & W R-2800 radials, and although the DC-6 was introduced more than a year after the Connie, the Douglas machine was an effective competitor. Indeed the DC-6's introduction started the colourful Lockheed-Douglas battle for piston-engined supremacy, which lasted until the late 'fifties when it was prematurely snuffed out by the advent of jetliners. By the time Model 049 production ended in mid-May 1947, 73 aircraft had been delivered in six versions to Pan Am (22), TWA (28), Air France (4), AOA (7), KLM (6), LAV (2) and BOAC (4). Some Model 049s were retrofitted with extra fuel capacity and became 149s, and four proposed Connie developments were never built, these being the 249 (a bomber version), the 349 (a long-range troop transport) and the 449 and 549.

The next Connie was the Model 649, which first flew in October 1946 as NX101A. Engagingly called the Gold Plate Connie, the 649 had 2,500 hp Wright R-3350-749C18BD1s giving 13.6 per cent more power than the 049's engines, and with its 94,000 lb take-off weight the 649 had a slightly better power-to-weight ratio than its predecessor. Another improvement was reversible propeller pitch for helping brake during the landing run, a feature which became virtually standard on modern post-war airliners and continues to this day in the form of reverse thrust on jetliners. Not surprisingly, with its more powerful engines, the 649's cruising speed rose from the 049-46's 313 mph at 20,000 ft to 327 mph. Although other airlines expressed interest in the 649, only Eastern Air Lines placed orders and 14 were delivered from May into July 1947. The improved 649A followed, with a 4,000 lb take-off weight increase (98,000 lb) and the few aircraft built went to Chicago & Southern Air Lines which, incidentally, merged with Delta Air Lines on 1 May 1953. Further improvements led to the Model 749, with the weight creeping up to 102,000 lb, and, for the first time, Curtiss Electric propellers were offered as alternatives to Hamilton Standard ones. Model 749s went to Aer Linte Teoranta (5), Air France (9), Air India (3), Eastern Air Lines (7), KLM (13), LAV (2), Pan Am (4), Qantas (4) and TWA (12). But more was to come, and the final civil Connie development was the 749A, with yet another weight increase, to 107,000 lb, and the 749A went to Air France (10), Air India (4), Avianca (2), KLM (7), SAA (4) and TWA (26). The USAF also received ten military 749s, nine being C-121As with large cargo-loading doors and reinforced floors, and the remaining machine being a VC-121B VIP aircraft. Also, the US Navy took two Airborne Early Warning (AEW) 749As as PO-1Ws, with large dorsal and ventral radomes.

In the meantime Douglas was updating the rival DC-6, and this machine was developed into the superb DC-6B with more powerful 2,400/2,500 hp Pratt & Whitney R-2800s, making its first flight on 2 February 1951. Something better than the

Connie was needed so Lockheed brought out the Super Constellation as a replacement, and Connie production ended in September 1951 after 233 had been built. With the same 123 ft span as the Connie, the Super Connie was stretched to a length of 116 ft 2 in from its predecessor's 95 ft 3 in. But the most significant change was in the engines, because it was intended to power the Super Constellation by a version of the Wright R-3350 with exhaust-driven power-recovery turbines, raising power to 3,250 hp and, incidentally, making the exhaust much quieter. However, these unusual Turbo Compound engines were not ready in time, and the first Super Connie version (Model 1049) had to make do with much lower-powered 2,700 hp R-3350-956C18CA1 engines. The only airlines to take non-Turbo Compound Super Connies were TWA, which received ten, and Eastern, which took 14 and put the type to work in December 1951. Different internal layouts were used by TWA and Eastern, with the latter opting for 88 passengers while TWA ranged from 52 to 75. But eventually 102 passengers were squeezed in with high-density seating! Subsequent Super Connie versions had their rightful Turbo Compound engines, the first being the Model 1049A, of which 72 went to the USAF as RC-121D AEW aircraft, while the USN took 142 WV-2s for AEW/ECM and eight WV-3s for weather reconnaissance. Next came the 1049B, and ten of them went to the USAF as RC-121C AEW aircraft and one as a VC-121E for Presidential use; also, 50 R7V-1 transports were delivered to the USN.

Neither the 1049A or B served with airlines, and the second civil Super Connie was the Model 1049C which entered service in June 1953, with KLM. Transatlantic flights started in August and, a major step forward, KLM's 1049Cs could fly non-stop eastbound from New York to Amsterdam, although prevailing winds on westbound flights made refuelling stops at Shannon or Prestwick necessary. Other 1049Cs went to Air France (10), Air India (2), Eastern (16), PIA (3), Qantas (3) and Trans-Canada Air Lines (5). Model 1049Ds were convertible cargo/passenger aircraft taking 104 passengers or 18 tons of freight and they had a take-off weight of 135,000 lb. Then came the 1049E passenger airliner which went to Air India (3), Avianca (3), Cubana (1), Iberia (3), KLM (4), LAV (2), Qantas (9) and TCA (3); also, the USAF took 33 1049Es as C-121C transports. Continuing competitive pressure from the Douglas DC-7 series forced Super Connie development, and the next version, the 1049G, entered service in 1955. Take-off weight was up to 137,500 lb, 60 first-class or 99 tourist-class passengers could be carried, and with its ability to cross the Atlantic non-stop in both directions under favourable weather conditions, the Super-G brought nearer the ultimate goal of routine non-stop transatlantic flights. The Super-G was the most-built civil Super Connie, and over 100 were sold to 17 customers including Air France, Air India, KLM, TWA, Lufthansa and others. The final civil Super Connie was the Model 1049H, a convertible cargo/passenger aircraft which entered service in October 1956; 53 were built.

Further development was needed to counter the Douglas DC-7C, the first airliner able to cross the Atlantic non-stop both ways regardless of weather conditions, and this impressive long-hauler started work with Pan Am on 1 June 1956. Lockheed replied with the Model 1649A Starliner, a refined Super Connie derivative with a new 150 ft span wing. This magnificent machine had other improvements such as increased fuel capacity and a lower propeller tip speed, and TWA started Starliner flights over the North Atlantic on 1 June 1957. But this was a full year after the start of the DC-7C's transatlantic services, and, even worse, was little more than a year before the arrival of jetliners — the Starliner never stood a chance. Few major airlines wanted to buy piston-engined machines at this late stage with jetliners just round the corner, and so the Starliner found only three customers — TWA, which bought 29 Starliners, Air France (10) and Lufthansa (4). The superb Starliner was the swansong of long-haul piston-engined airliners, a breed struck down prematurely by the Boeing 707 and Douglas DC-8 jetliners, but 856 Connies and derivatives were built, underlining the success they enjoyed during their heyday.

First flown on 9 January 1943 as a C-69 military transport (NX25600), the Lockheed Constellation could reach the speeds of some of its fighter contemporaries thanks to its powerful 2,200 hp Wright R-3350-35 engines. When WW2 ended, the Connie started a new golden age of airline flying, by making possible non-stop long-distance travel unknown during the 'thirties except by specially-built or modified aircraft. It was the Constellation, along with the Douglas DC-6 and Boeing Stratocruiser, which made regular and frequent commercial transatlantic flying quite normal after WW2. The picture on the left shows a C-69 flying over mountainous territory.

B. Robertson

American Overseas Airlines' N90925 *Flagship America* is parked in the company of a Douglas DC-3 behind and a Douglas DC-6, its major competitor, to its left. This Connie joined AOA in May 1946 but the airline was bought by its competitor Pan American on 25 September 1950, so Pan Am acquired AOA's seven Connie Model 049s, renaming them *Clippers Jupiter Rex, Mount Vernon, Golden Rule, Lafayette, Courier, Ocean Herald* and (particularly pleasant) *Wings of the Morning* — corresponding registrations being N90921 to '27. When Delta Airlines found itself short of aircraft in 1956 it bought four ageing 049s from Pan Am, one of them being the aircraft shown here.

MAP

MAP

During the late 'forties Great Britain had a large and experienced aircraft industry which had been vastly expanded during WW2, and would therefore have been expected to meet British airlines' needs. Unfortunately British industry failed to produce a decent long haul airliner, and British Overseas Airways Corporation had to buy American; so between 1946 and 1955 BOAC bought eight Connie Model 049s and 17 Model 749As. Shown here is 049 G-AKCE *Bedford* which joined BOAC in March 1948 and was eventually sold to Capital Airlines in the USA in June 1955 as N2741A.

Nose livery and the legend AA under the starboard wing identify the Connie above as one of the seven Model 049s bought new by American Overseas Airlines. The photographer is about to be well and truly blasted by the noise, and even at this early stage of take-off the mainwheels are already retracted while the nosewheel rapidly follows suit. Flaps are set to the take-off angle, the landing angle being much greater than this, and engine cooling-air exit flaps are visible. Other things to note are the de-icer boots on the wing leading edges, seen as black cappings outboard of the inner engines going right up to the wing tips; these rubber boots pulsated to dislodge ice forming on the leading edges (the most likely place). AOA was a very early user of Constellations, which put it into something of a select position, ie, it was an airline with Connies! Actually, the Connie had competition from the excellent Douglas DC-6 which turned out to be a formidable rival, selling well, but it did not enter service until over a year after the Connie. However the DC-6 lacked the magic of the Constellation which had a certain *cachet* thanks to its beautiful lines, its celestial name (DC-6 sounds clinical) and the fact that it was the first long-haul airliner in the post-war mould — ie, four aircooled radials of over 2,000 hp, a transatlantic range and a cruise of up to around 300 mph.

On the left, an El Al Constellation Model 049 (4X-AKB) taxis out at what is certainly a British airport, as shown by the Jaguar car, Bedford or Austin lorry and the British look of the bus' rear which can just be seen — Heathrow perhaps? No doubt this machine is moving out for take-off because the flaps are at their take-off setting. Note the Connie's considerable height, the twin main and nosewheels, and the undercarriage doors which close when the wheels are up to cover the gap. This aircraft is a very early Constellation and, starting off as a C-69 with the USAAF in early 1945, it joined El Al in March 1951.

M. J. Hooks

Perhaps no airliner has ever been so graceful as the Connie, and modern jetliners look bland by comparison. However the Connie's eel-shaped fuselage had its price, because stretching, which is often the prerogative of successful aircraft, would have been easier if the Constellation had had a parallel-sided fuselage. The rival Douglas DC-6 had this feature and it was only necessary to insert a straight section into its fuselage to increase capacity, but Lockheed would have had to blend in a new curved section. This may account for the fact that the Constellation remained unstretched until the Super Constellation was introduced, whereas the DC-6 was stretched into the DC-6A and DC-6B. The machine shown above, N88861 *Clipper Winged Arrow*, was an 049 Constellation, and Pan American bought 22 of them, all being delivered from January to May 1946 as N88831 to '3, N88836 to '8, N88845 to '50, N88855 to '62, N88865 and '68. Pan Am's aircraft had names starting with the word *Clipper*, some of the names being truly delightful. For instance there was *Clipper Golden Fleece* (N88868), the Golden Fleece of Greek legend taken from the ram which bore Phrixus through the air to Colchis! Another gem was *Clipper Empress of the Skies* (N88858), a name reflecting the majesty of the Constellation, while *Clipper Unity* (N88857) had a tranquillity quite out of keeping with the Connie's totally untranquil 2,200 hp Wright Cyclone 18s!

B. Robertson

The improved Connie Model 649 was introduced in October 1946 with Wright R-3350-749C18BD1 engines increasing power from 2,200 hp to 2,500 hp per engine; reversible-pitch propellers were also fitted. Then came the 649A with more take-off weight, followed by the 749 with yet another increase in weight from 98,000 lb to 102,000 lb; a further development was the 749A with a five per cent weight increase to 107,000 lb. The 749 could cruise at up to 327 mph, a formidable performance which WW2 RAF bomber crews would have more than welcomed only a few years before in their Lancasters, Halifaxes and Stirlings, none of which had even a maximum speed of 300 mph, let alone cruise. Aircraft are usually developed to carry ever more loads, and the Constellation, unusual among airliners then and since, was sometimes fitted with a removable ventral pannier (the Speedpak) which was first carried by the 649. The aircraft shown to the left is a 749 belonging to TWA, a major Connie operator.

The Speedpak ventral pannier added about 400 ft³ of cargo volume to the aircraft's existing 445 ft³, allowing an extra load of up to 8,200 lb. Fitted with small rubber-tyred wheels for ground handling, the Speedpak was 33 ft 4 in long, 7 ft wide, 3 ft deep and it weighed 1,836 lb; one of its benefits was to allow loading on the ground independently of its carrier aircraft, which avoided expensive aircraft waiting time. Built-in electric lift motors raised the Speedpak up against the underside of the Connie and it was claimed that a loaded Speedpak could be raised and locked in place in two minutes. Despite its additional bulk and weight, it reduced speed by only about 10 mph, but naturally it cut down range when the Connie was heavily-loaded with passengers and cargo. Some of the Speedpak's users included KLM, Eastern and TWA.

B. Robertson

NC86520 *Clipper America*, below, was one of five Connie 749s used by Pan Am, the others being N86527 *Clipper Glory of the Skies* (very much in keeping with the Connie's place in the airliner pecking order in 1947), *Clipper Sovereign of the Sky* (N86528), *Clipper Monarch of the Skies* (N86530), and *Clipper Romance of the Skies* (N86529) — and yes, airline flying *was* romantic 45 years ago. Readers with an eye for detail might try to spot the difference between this machine and Pan Am's 049 shown on the previous page. The newer machine has one less side window, cockpit roof glazing, and there is only one exhaust pipe to be seen on the outboard side of each engine at the rear of the cowling, while the earlier machine seems to have two. Pan Am's Model 749s were all delivered in June 1947, and had the same 123 ft span and 95 ft 3 in length as the 049, but the 749 had 2,500 hp Wright R-3550-749C18BD1s instead of the 2,200 hp BA3s of the earlier Connies and its gross weight of 102,000 lb was 18 per cent more than the 049-46's 86,250 lb. Pan Am inaugurated 749 services with NC86520 shown here, and this Connie distinguished herself by flying Pan Am's first round-the-world service from New York and back in June 1947. Leased to Pan Am, this machine was used by the airline for less than a year before being returned to Lockheed late in 1947; it then went to Aerovias Guest and eventually joined Air France as F-BAZR in January 1949.

MAP

MAP

It is obvious that the scene above is British from the houses and the four motorcars, which car connoisseurs will recognise as, from left to right, Austin 7, Austin 7, Ford 8 (to the left of the Connie's starboard undercarriage leg) and a Ford 8 or Anglia by the port leg. It might be Heathrow. The portable fire extinguisher looks ominous, but it was normal to have a man standing by with a fire extinguisher under the engine at start-up — nobody bothers these days! Despite being an avid airliner watcher during the post-war piston years I never once saw an engine catch fire during start-up, although the Douglas DC-4, in particular, sometimes spat back with a flash of flame through a carburettor air intake. Luckily the intake was forward-facing so the passengers never saw it! On start-up thick smoke used to pour briefly from the exhausts, and sometimes the odd flame. Even in repose, with its massive undercarriage doors and drooping elevators, Qantas' Model 749 VH-EAC *Harry Hawker* looks supremely elegant. Note the small drooping flap visible under No 1 (port outer) engine; this is a shutter to control the airflow through the oil cooler, whose scoop can be seen at the front of the engine underneath the propeller.

MAP

PH-TEP *Pontianak* was one of 20 749/749As operated by KLM and the machine to the right was delivered in 1947. Flap setting suggests that this Flying Dutchman is taxying out for take-off, so some of the passengers will be excited and nervous. Excited, because airline flying was still rare enough to be a special experience. And nervous? Because, quite simply, of the possibility of the unthinkable happening at take-off. Also, compared to a modern jetliner the take-off noise was pretty sobering, and this Connie has jet-stack exhausts in which short pipes from the engine gave some jet thrust at the expense of a great deal of noise. At take-off they sounded rather like millions of small stones running down an enormous chute!

M. J. Hooks

Continually improving the Connie to keep abreast of competition from Douglas, with its excellent DC-6 series, Lockheed introduced the 749A in 1949 with revised landing gear allowing a take-off weight of 107,000 lb. VT-DAS *Himalayan Princess,* shown above, can be identified as a 749A by the lack of forward-facing air intakes above the cowlings; she was one of four used by Air India and was delivered in January 1950, being eventually traded in for a Super Constellation. Note the absence of propeller spinners. The pointed hubs, containing the pitch-change mechanism, show that these are Curtiss Electric propellers instead of Hamilton Standard ones — other give-aways are the blade-root cuffs. A major difference between Hamilton props and the Curtiss ones used on some 749s and 749As lay in the way the pitch was changed, hydraulically on the former and elctrically on the latter. Like the KLM machine on the previous page, this Connie has noisy jet-stack exhausts. Manufacturers often tried to ensure that exhausts discharged outboard of the engines or under the wing to avoid direct exposure to passengers, but Lockheed did not do this with the jet-stack system so passengers just had to grin and bear the noise from the inner pipes facing the cabin! one can almost hear the menacing crackling snarl from the exhausts as the machine taxis by. Air-India also received three 749s, VT-CQP, 'R and 'S, and all its Connies were called *Princesses,* with graceful names befitting the Connie's appearance, and in addition to the machine shown the 749A fleet comprised VT-DEO *Bengal Princess,* VT-DEP *Kashmir Princess* and VT-DAR *Maratha Princess.* Unfortunately a bomb was planted in *Kashmir Princess* and exploded in flight, but the Connie managed to ditch. Far worse was the tragic fate of the Model 749 VT-CQP *Malabar Princess* which, flying in a heavy snowstorm and high winds on 3 November 1950, flew into the Mont Blanc in France, near the Italian border. There were no survivors. Four Connie 749s were delivered to Qantas in October 1947, one of them being VH-EAA shown here and named after the famous Australian Ross Smith who, with his brother Keith, flew from Great Britain to Australia in 1919 leaving Hounslow on 12 November and arriving in Darwin on 10 December. They flew in a Vickers Vimy bomber, a rather different machine to the Connie, the latter having three and a half times more power in one engine than the combined urge of the Vimy's two Rolls-Royce Eagles! That a Commonwealth country like Australia should have had to buy American aircraft, despite trading attitudes of 45 years ago, simply

highlighted the inability of British industry to produce viable long-haul aircraft. This machine made the first Qantas proving flight from Sydney to London, in November 1947, and was converted to a 749A during its time with Qantas in which form it is seen here. Like the Air-India Connie above, *Ross Smith* had noisy jet-stack exhausts which were none too popular with passengers, who probably failed to appreciate the use of jet thrust to improve economy and keep their fares down! This machine stayed in the Commonwealth "family" for some years because it was sold to BOAC as G-ANUP *Branksome* in February 1955. It came to an extraordinary end when, no longer with BOAC, it tried to overshoot with superchargers in High Gear, used for boosting the engines at high altitudes in less dense air and definitely not to be used at low altitudes! The consequent over-boosting damaged the engines and caused failure.

Qantas

MAP

Crossing the Atlantic during the late 'forties in a piston-engined airliner made for a long and tiring trip. And because aircraft such as the Connie could not fly London-New York non-stop, an intermediate landing was made at Shannon, in the west of the Irish Republic, to top up with fuel for the long haul across the pond; sometimes another stop was made at Gander in Newfoundland. It was said that Irish coffee was concocted to fortify passengers (but presumably not crews) for the long transatlantic haul! This shot might be Shannon Airport, some of the vehicles are British and the eagle-eyed and knowledgeable will spot three Austins and a Standard. In the background, to the right, is a sign for British South American Airways, whose Director and General Manager was Air Vice-Marshall D.C.T. Bennett of RAF WW2 Pathfinder fame. The Connie is EI-ADD *Saint Kevin*, and was one of five 749s delivered in 1947 to Aerlinte Eireann Teoranta, and three of the Connies made a formation delivery flight across the Atlantic! However Aerlinte's Connies were never used in the manner intended because, although bought for the North Atlantic service, the plan was cancelled and the five Connies were sold to BOAC which received them in June 1948. This machine went to BOAC as G-ALAN *Beaufort* (all BOAC's Connies had names beginning with B), and during its time with BOAC the machine was converted to 749A standards. As shown here the aircraft does not have jet-stack exhausts, instead the exhaust is collected in a manifold and an exhaust pipe is just visible under each starboard engine. The flaps about halfway between the wing leading edge and the front of each engine are cooling air exits for the 2,500 hp Wright Cyclone 18 engines.

MAP

There is no mistaking the location of this Air France Connie 749 in the bottom picture as Heathrow, probably in the 'fifties — note the two Bedford Dormobile-based vehicles in attendance. Air France bought four Model 049s, nine 749s and ten 749As, and the 749 shown (F-BAZJ) was originally started as a 649, emerged as a 749 and was converted to a 749A during its time with Air France! It was delivered in June 1947 and was taken out of service in late 1969 serving with the Armée de l'Air. Like some of the machines shown on previous pages, this Constellation has jet-stack exhausts, which I remember so well for their excruciating note at take-off from Geneva Airport during the 'fifties. These jet-stacks added a few mph to the cruising speed and improved climb, and Air France was a particular exponent of the jet-stack system.

Despite the Connie's excellence and the *cachet* associated with flying in it, Lockheed faced stiff competition from the Douglas DC-6, which, like the Connie, was being improved and updated. The Douglas DC-6B entered service with American Airlines on 29 April 1951 with an even better combination of performance, reliability and economics than its DC-6 parent. It was also faster than the Constellation, and production of the Connie ended in September 1951 after 233 had been built. This was a good production run but the fact remains that the DC-6B stayed in production until 1958, and 704 of the DC-6 in all its forms were built which was over three times as many as the Connie in all its forms. The magnificent DC-6B played its part in forcing Lockheed to stretch the Constellation into the Super Constellation, and the prototype Connie was rebuilt as the Super Connie

MAP

prototype in 1950. But the first aircraft built from the ground up as a Super Connie was N6201C, the Model 1049, with its fuselage stretched from the Connie's 95 ft 3 in to 113 ft 7 in but with the same 123 ft span as the Connie. Wright Aeronautical was about to bring out a remarkable Turbo Compound version of the Wright Cyclone 18, fitted with exhaust-driven turbines to increase power from 2,700 hp to 3,250 hp. But the TC engine was not ready in time for the Super Connie's debut and so the much less powerful 2,700 Wright R-3550-956C18CA1s had to be fitted instead. With a take-off weight of 120,000 lb the new Super Connie 1049 was 12 per cent heavier than the 749A, but it initially had only eight per cent more power until 2,800 hp 975C18CB1s were fitted which restored the power-to-weight ratio. The 1049 went into airline service with its stop-gap non-Turbo Compound engines, and the machine shown in the top picture joined Eastern Airlines which bought 14 Model 1049s, introducing the type in December 1951. The Eastern Airlines machine shown in the bottom picture is N6202C, which became part of the company's famous "Great Silver Fleet", and the reader may have noticed the jet-stack exhausts — noisy as ever — which were fitted to early 1049s before the Turbo Compound engine became available. Not surprisingly the Super Connie shares a strong similarity to its Connie parent, and despite any difficulties Lockheed successfully stretched the aircraft's curved fuselage, although perhaps more expensively than would have been the case on a parallel-sided fuselage like the DC-6's. If anything, the stretched fuselage enhances the Super Constellation's lines, yet the converse was true when the Douglas DC-6 was stretched into the DC-7. The cockpit glazing and roof are quite different to those on the Connie, with the fuselage flattening out over the Super Connie's cockpit.

B. Robertson

B. Robertson

The absence of a white cabin top, so fashionable on airliners later on and which was said to keep the cabin cool in the sun, makes Eastern Air Lines' Super Connie look austere and unfinished. This machine, incidentally, was withdrawn from service in 1968 by Eastern and went through a number of owners before being scrapped after a forced landing in 1973. Other Model 1049s in Eastern's "Great Silver Fleet" were N6203C to '14C, and the first 14 1049s all went to Eastern. N6212C had a bizarre ending when one of Eastern's Douglas DC-7Bs, with failed steering gear, taxied straight into the Super Connie! Leaking fuel from a damaged tank on the Super Connie caught fire and both aircraft were wrecked, luckily without injury. Also, N6213C had a narrow escape in service with Eastern in May 1962 when the nosewheel collapsed on landing at New York after a flight from Washington — nobody was injured and the machine was repaired. Yet another mishap finished off the final Eastern 1049 (N6214C) when it caught fire after an emergency landing due to engine problems during a flight from Seattle to Chicago — all on board survived. With the introduction of its glamourous new Super Constellation, Lockheed evidently thought it a good time to remind the world of the firm's illustrious past and promising future, with this family gathering shown above. The machine in the foreground is N6202C, the aircraft in Eastern Air Lines livery on the previous page.

It is always interesting to see an aircraft in the social, economic and environmental context of its time, and in this respect the cars behind N6202C are of interest; the white one looks like a product of the Chrysler Corporation. Incidentally there was far less environmental concern than there is now and green was just a colour in those days! The second aircraft in the line-up is a Model 749A (F-BBDV) which was the final Connie, and it went to Air France in September 1951; it is just possible to spot its lack of propeller spinners. The smaller twin-finned trio are all pre-war machines and from left to right are Model 10 Electra (1934), Model 12 Electra Junior (1936) and Model 18 Lodestar (1939), all of which paved the way for the Connie by giving Lockheed experience in the design and production of high-performance all-metal cantilever monoplanes. The Model 10 was Lockheed's first all-metal machine, indeed it was Lockheed's first twin, and was commercially very successful, establishing the firm in the all-metal field alongside Douglas, Boeing and others. British Airways, instead of buying British aircraft as might have been expected in the 'thirties, bought Lockheed 10s. But then there was nothing on this side of the pond to match the Model 10 and BA was not the national flag carrier that it is now (that was Imperial Airways), so it could buy what it liked. The second twin-fin machine, the Model 12, was a smaller version of the Model 10, but with the same engines and a more sprightly performance.

Again, the Model 12 was a very successful aircraft and sold well mainly to corporate and private owners. It had absolutely no competitor in Europe, which was one of the reasons why the American aircraft industry became dominant in the civil field, a position which the Connie and Super Connie did much to maintain. The rearmost twin-fin machine, the Lodestar, was produced to meet the need for greater capacity and was developed from the smaller Lockheed 14. Surprisingly no Model 14 is shown here, because when this picture was taken, probably in 1951, it should have been possible to get hold of one. Finally, the enormous double-deck machine in the background is the Constitution, not one of Lockheed's proudest achievements and only two were built, both going to the USN as heavy transports.

B. Robertson

Apart from Eastern Air Lines, TWA was the only other user of the original Super Constellation 1049, which had to manage without the Wright Turbo Compound engines intended for it. N6902C was one of ten delivered to TWA, and the aircraft's name, *Star of the Seine*, is stencilled in English and French just below the cockpit glazing and above the rear window. Indeed all TWA's 1049s were *Stars*, and other aircraft included *Stars of the Thames* (N6901C), *Tiber, Ganges, Rhone* and *Rhine*, as well as *Stars of Sicily, Britain, Tipperary* and *Frankfurt* (N6903C to '10C), and they all went into TWA ownership between May and September 1952. The TWA machine shown here was lost on 30 June 1956 in a mid-air collision over the Grand Canyon with a United Air Lines Douglas DC-7, a rival type, with the loss of all on board. Unusually, *Star of the Ganges*, suffered a double engine failure on the starboard wing in December 1952, making an emergency landing in which the machine was damaged, but later repaired. By the 'fifties aero engines were extremely reliable and a double engine failure was a very rare event, the cause in this case being a gear train failure in the cam drive of *both* engines! Fortunately there were no injuries, but *Star of Sicily* was involved in a much more serious accident on 16 December 1960, in a mid-air collision with a United Douglas DC-8 jetliner over Staten Island. There were no survivors. A civil Super Constellation with its rightful Turbo Compound engines appeared as the Model 1049C, first flying on 17 February 1953; this was not the first Turbo

MAP

Compound-powered Super Connie because both the 1049A and 1049B had these impressive engines, but these were military aircraft. Despite the obvious advantages of compounding, the Wright Turbo Compound was remarkable in that it was the only production aero engine with exhaust-driven power recovery turbines. About 550 hp was simply going to waste in the exhaust, and the reasoning behind the Turbo Compound was very simple. Why waste all that exhaust energy (which also makes the exhaust so noisy) when it could be used to drive turbines whose power could be fed back to the crankshaft? Even if jetliners had not made long-haul piston-engined airliners prematurely obsolete then turboprops would probably have done so, but if piston engines had been given a longer innings it is quite possible that compounding

B. Robertson

would have been taken up by other manufacturers in the interests of economy. As it was, Wright Aeronautical was the only firm to pursue the idea into production, and in so doing it failed to develop a replacement engine at a time when piston engines were growing obsolete, in the mistaken belief that the Turbo Compound would carry it through. And so it did for a few golden years during which it sold well. Sometimes advertised as the "Non-Stop" engine because of the great ranges of some of the aircraft it powered, the Turbo Compound had three exhaust-driven turbines geared to the crankshaft through hydraulic couplings, each turbine being supplied by six of the engine's 18 cylinders. Thanks to the energy extracted from the exhaust by the turbines, not only was power increased by 20 per cent from 2,700 hp to 3,250 hp, but the exhaust became astonishingly quiet. However, the reduction in exhaust noise was offset by the vast and extraordinary increase in propeller noise, and the eerie wail of the Super Connie at take-off is something that I will never forget. Even during cruise the noise was very noticeable, a kind of deep, ponderous rasp which was instantly recognisable. KLM started transatlantic services with Super Connie 1049Cs in August 1953, and shown on the left is the first 1049C to serve with the Dutch airline, PH-TFP *Elektron*. Although the spectre of nuclear war was as uppermost in people's minds then as it was in more recent times until Glasnost, atomic energy was also seen as a promising new form of power. The environmental problems seen to be so important these days did not worry the public during the 'fifties, and it may have been in this new spirit of cheap and abundant power for all that KLM called its nine Super Connie 1049Cs *Atoom* (PH-TPF), shown here being christened, *Elektron* (PH-TFR), *Proton* (PH-TFS), *Neutron* (PH-TFT), *Photon* (PH-TFU), *Meson* (PH-TFV), *Deuteron* (PH-TFW), *Nucleon* (PH-TFX) and *Triton* (PH-TFY). Some of these names might sound spine-chilling today!

B. Robertson

With the introduction of the Super Connie 1049C KLM was able to fly non-stop eastbound from New York to Amsterdam, while for the westbound trip prevailing winds made a stop at Shannon or Prestwick necessary for fuel. Despite the latter constraint this was a major step forward in transatlantic flying, where the ultimate goal was to fly regularly non-stop between major European centres and New York in either direction. KLM's 1049Cs started on transatlantic routes two years before the rival Douglas DC-7B, which had to wait until 13 June 1955 before it, too, started crossing the Atlantic when Pan Am put it on non-stop New York-London services; so Lockheed was in pole position for a time. Despite its 1955 transatlantic debut, the DC-7 started work with American Airlines in November 1953, and like the Super Connie 1049C the DC-7B could not fly the Atlantic non-stop in the westbound direction. Also powered by Turbo Compound engines — there being no other competitive piston engines — the Douglas DC-7 series had four-blade propellers instead of the Super Connie's three, and with a lower tip speed they were virtually inaudible. Indeed, the DC-7's ideal combination of quiet exhausts and *sotto voce* propellers made it something of a quiet giant even at take-off, quite unlike its cacophonous smaller brother, the DC-6B. Incidentally, Lockheed never fitted four-blade propellers to its piston-powered Super Connies, and chose to keep the rasping three-bladers instead!

Apart from the nine 1049Cs ordered by KLM, others went to Air France (10), Air India (2), Eastern (16), Pakistan International Airlines (3), Qantas (3) and Trans-Canada Air Lines (5). Qantas' machine VH-EAG, shown here, is aptly named *Southern Constellation* and it inaugurated Qantas' Super Connie services in May 1954, flying from Sydney to San Francisco. It was traded in to Boeing for 707s in May 1963 — a symbolic transaction in a way, because it was the Boeing 707 and Douglas DC-8 jetliners which prematurely

B. Robertson

B. Robertson

ended the careers of large long-haul piston-powered airliners, like the Super Connie, when they still had years of life left in them. The other two Qantas machines, *Southern Sky* (VH-EAH) and *Southern Sun* (VH-EAI), were both traded in for turboprop Lockheed Electras after serving for only five years. *Southern Sky* inaugurated Qantas' Sydney-London Super Connie service in August 1954, while *Southern Sun* was converted to a 1049E later on. Now have a look at F-BGNA, the Air France 1049C on the left. The flap angle shows that it is taking off rather than landing, and, of course, the camera crew will shortly be in the plane of the propellers and will either enjoy, or cringe from, this noisy experience! The slight cutaways just behind each engine cowling are to clear the exhaust pipes, which emitted flames at night! Indeed I remember a Super Connie taking off in daylight from Geneva during the 'fifties with a small but persistent and worrying jet of exhaust flame coming from one of the engines. This was certainly not normal by day but I supposed that the engine instruments read normally, and take-off was continued — perhaps the Super Connie was too near the end of the runway to do anything else! Anyway, it climbed away normally.

Under competitive pressure from Douglas, Lockheed updated the Super Constellation and the next model was the 1049D, of which four were built, all entering service with Seaboard & Western Airlines as convertible freight and passenger aircraft carrying 18 tons of freight or 104 passengers. By this time weight had crept up to 135,000 lb from the 1049C's 133,000 lb. Then came the Model 1049E, a passenger aircraft, 28 of which were delivered to various airlines. But greater range was needed and so the Model 1049G was introduced, entering airline service in 1955. This machine could cross the Atlantic non-stop in both directions under favourable weather conditions and was often known as the Super-G, some of them being fitted with wingtip tanks. The machine about to give the photographer's ears a good battering is CS-TLA, a 1049G of Transportes Aereos Portugueses (TAP) taking off, while above it is PIA's AP-AFQ, one of three Model 1049Cs owned by the Pakistani airline. But back to the Super-G. It was the most prolific civil version of the Super Connie, and this, as well as its distinctive nickname, helped ensure it was the best known. Take-off weight was up to 137,500 lb and the Super-G carried 60 first-class or 99 tourist-class passengers; over 100 went to 17 customers worldwide.

B. Robertson

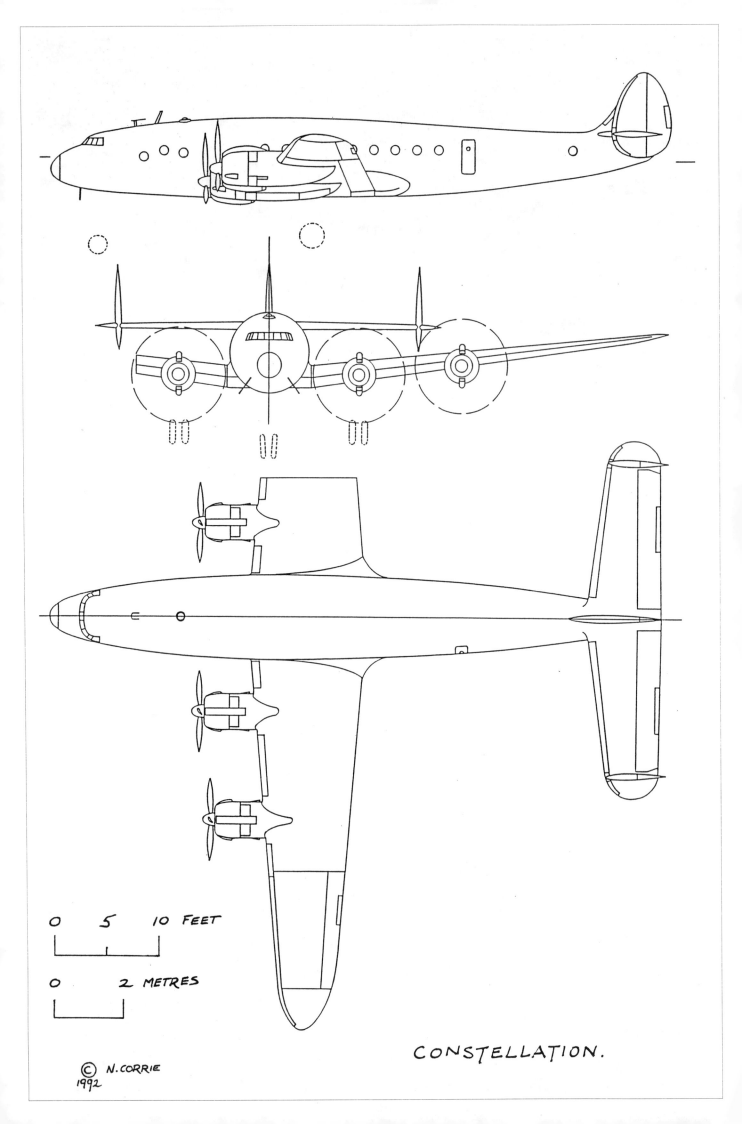

0   5   10 FEET

0   2 METRES

© N.CORRIE
1992

CONSTELLATION.

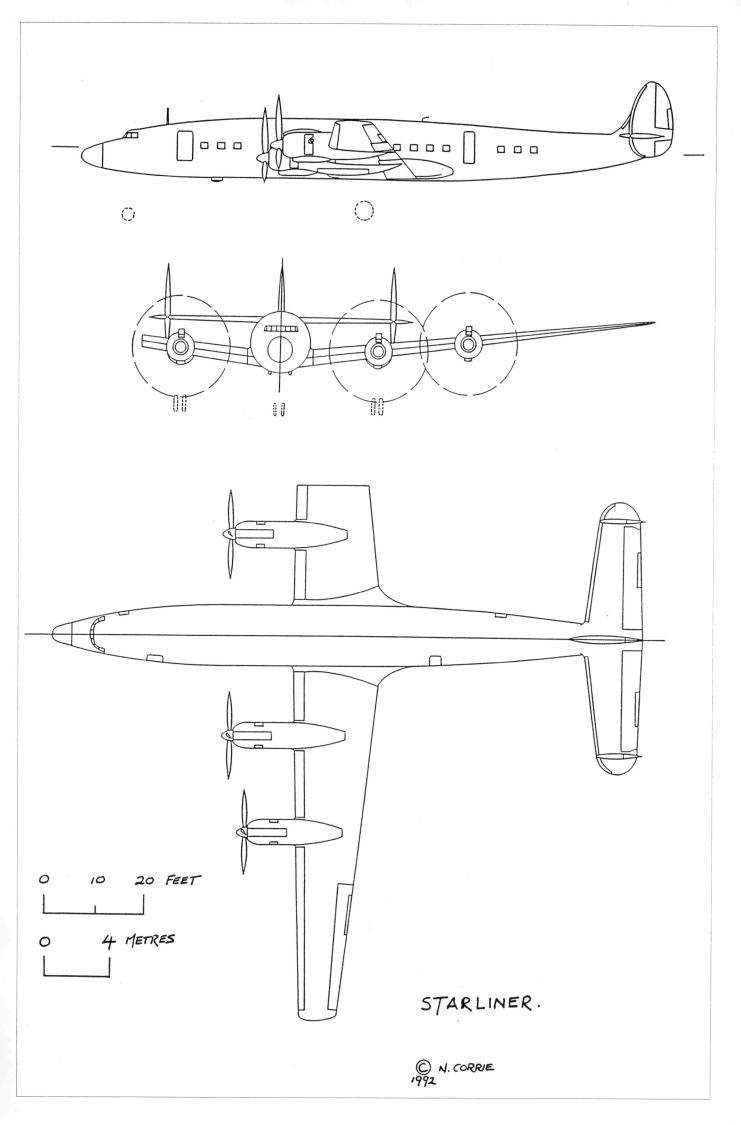

O   10   20 FEET

O   4 METRES

STARLINER.

© N. CORRIE
1992

Qantas

TWA's Super-Gs were the first of the breed to carry nose-mounted storm-warning radar, and this equipment became evident on some airliners from the mid-fifties onwards; the Qantas machine VH-EAD shown here is fitted with it. *Southern Dawn* was one of nine 1049Es which went to Qantas but it was converted to a 'G later on and delivered in late 1955, yet less than five years later it was traded in for Lockheed Electras, in March 1960. Although Electras were turboprop machines, this Super Connie's short innings is a reminder that the introduction of jetliners (excepting the short-lived Comet operations between 1952 and '54) was only three years after *Southern Dawn's* delivery. And no matter how excellent and magnificent Super Connies were, there was no way that piston-engined airliners could compete with jetliners, so front-ranking airlines had no choice but to replace their relatively young piston machines with Boeing 707s and Douglas DC-8s. It was sad to see these fine piston-powered aircraft so prematurely displaced, but it had to be and with their demise came the end of an era. But this is to jump ahead. Trans Canada Air Lines was formed in 1937 and started operations with Lockheed 10 Electras between Vancouver and Seattle. The Model 10 was Lockheed's first all-metal twin, a machine in the modern mould which can be said to have been in concept a kinsman of the Connie. Two Super-Gs were delivered to TCA in 1956 and another two later on, enabling the company to increase its North Atlantic flights among others. TCA became Air Canada in June 1964, having stopped using Super Connies in 1963 when they were displaced by jetliners after a relatively short time. The TCA Super-G shown on the next page has no storm warning radar and is taxying on a wet day at Heathrow, but how can it be identified as a Super-G? It says so on the centre fin!

Contemporary machines in the background include a Pan Am DC-7, as well as two other machines which only the eagle-eyed will be able to identify! Note, also, the KLM building and the BOAC bus, as well as the ground staff member in regulation white overalls. The horizontal bar on the TCA machine just above the twin nosewheels carries two hydraulic steering cylinders, because the Super-G, like other large aircraft, needed power steering. Many other things can be noted about this Super Connie, for instance the two windscreen wipers suspended from the top of the windscreen, the afterbody behind each propeller spinner and the split undercarriage doors to cover the mainwheels when retracted, as well as the nosewheel doors. Incidentally, on some airliners the doors would close behind the wheels once they were down to reduce drag. The next picture shows Air France Super-G F-BHBB which was one of 14 to go to the French airline. No radar nose is fitted to this machine, which went to Air France in July 1955 and served with the airline for the next 13 years, being eventually scrapped in 1973. Other Air France Super-Gs included F-BHBA, F-BHBC to 'J and F-BHMI to 'L. Although an extremely large aircraft for its time, the Super Connie would be dwarfed by a modern-day Boeing 747, the latter being about twice the size. Pan Am and TWA competed with one another on the North Atlantic run, Pan Am with its Douglas DC-7Bs introduced on 13 June 1955, and TWA with its Super Connie 1049Gs introduced two months later on on 1 November 1955. The Super-G's engines were upgraded to the new Wright 972TC18DA3 Turbo Compounds which, with a take-off rating of 3,250 hp like previous Turbo Compounds, nevertheless had a METO power (Maximum Except Take-Off) increase of 100 hp to 2,750 hp (nearly four per cent) and it was this which allowed the gross weight to be raised to 137,500 lb; 3,400 hp EA series engines were fitted later on. The Super-G had various improvements, including better soundproofing, but it still had the horribly (or delightfully) noisy propellers of its predecessors. The relentless competitive pressure from Douglas continued, and Douglas introduced the DC-7C Seven Seas which was the first airliner able to cross the Atlantic non-stop, in either direction, against prevailing winds. Advantage, then, to Douglas and Pan Am when Pan Am put the aptly-named Seven Seas to work on 1 June 1956. Yet, sadly, it was only too soon before the arrival of jetliners to make this titanic battle between the Super Connie and the DC-7 utterly irrelevant. The pinnacle of piston-engined airliner design had almost been reached and the golden era was coming to its finale, so let us have a quick look at the unusual Turbo

Compound engine which made these magnificent aircraft possible. Despite its complexity and its use of technology shunned by other manufacturers, this engine worked satisfactorily in service, but it did give some maintenance headaches to certain airlines. The most interesting feature of this engine was, of course, its exhaust-driven power-recovery turbine system. Engineers will know that there are basically two types of turbine, namely the reaction turbine and the impulse turbine. In both types torque is obtained by trading gas pressure for gas velocity, normally called the pressure drop across the turbine, and on the impulse turbine all the gas pressure is traded for velocity before entering the turbine rotor. But on the reaction turbine the gas pressure drop takes place in the rotor itself. The reaction turbine is used in gas turbine engines and is more efficient because of the lower gas velocities through the rotor, causing lower frictional losses — but the Turbo Compound had impulse turbines. Engine RPM at take-off was 2,900, while the three turbines spun at 19,000 RPM, and not only this, because the turbines worked in the hot exhaust gases — an appalling environment so not surprisingly the turbines had to be aircooled. Reduction gearing and shock-absorbing hydraulic couplings were used to connect the turbines to the crankshaft.

B. Robertson

A journey safely completed, above, in TAP's Super-G CS-TLA. This is probably Heathrow and it could be any time between 1957, when this Super Connie started flying to Heathrow, and 1967 when it was withdrawn, but the elderly-looking Mobiloil vehicle on the left suggests the 'fifties. Indeed this shot is rich in the vehicles and dress of the era when ladies and gentlemen wore hats, and what a treat for vehicle spotters because among the disembarking passengers are a Ford Thames light van, a Bedford van further to the right and, very unusual, a Vespa motor scooter with a sidecar; the latter has Shell-Mex and BP Ltd on the front. People were sometimes taken to and from the aircraft by coach in those days but it was not unusual to walk, indeed at Geneva I always had to walk come rain or shine because the aircraft were parked near the passenger buildings. The fin of a Lufthansa Convair can be seen just below the Super Connie's

MAP

M. J. Hooks

rear fuselage, and note that there is a rudder on all three of the Connie's fins, continuing below the tailplane on the outboard fins. Again, one is drawn to remark on the graceful lines of Lockheed's piston-engined masterpiece, which still looks magnificent even when surrounded by people and airport paraphenalia. The machine underneath TAP's Super Connie is a Model 1049H of the American Flying Tiger Line, a machine known as the Super-H or Husky, and this one has the extended nose housing storm-warning radar. The 1049H was the final civil Super Connie and was a convertible passenger/cargo aircraft, 53 of them being produced.

Like the Super-G, the 1049H could also be had with wingtip tanks. Making its first flight in September 1956, it started off with the same 137,500 lb maximum take-off weight as the Super-G, and it was given a strengthened floor among other mods. Early on during the Super-H's life the more powerful 988TC18EA3 Turbo Compound was fitted giving 3,400 hp for take-off, an increase of about 4½ per cent, and the Flying Tiger Line was the first Super-H operator to use this engine which allowed an increase in take-off weight to 140,000 lb. The machine shown was delivered to the Flying Tiger Line in March 1957, only about 18 months before long-haul jetliners started operating. However it was not jetliners which put N6915C out of business prematurely, but a tragedy just after take-off from San Francisco when the machine crashed and caught fire on Christmas Eve 1964 on a cargo flight, with the loss of all the crew. Other operators receiving this ultimate civil version of the Super Connie were Qantas (2), Seaboard & Western (5), Flying Tigers (13), California Eastern Aviation (5), Dollar Airlines (1), Resort Airlines (2), National Airlines (4), Slick Airways (3), Aerovias REAL (4), PIA (2), TWA (4), KLM (3) and TCA (2). In addition the Air Finance Corporation bought three for leasing. If the Lufthansa Super-Gs shown above in the hangar had been warplanes, then no doubt a suitable caption would have said "The sinews of war . . ." but airliners never seem to be called the sinews of peace! Super Connies stretch as far as the eye can see in this enormous maintenance hangar at Hamburg, but closer inspection will show that the furthest aircraft is not a Super Connie, but a Convair medium-haul twin, also used by Lufthansa. The notice on the far wall forbids smoking — *Rauchen Verboten!*

This hangar was completed in 1954 and could accommodate ten twin-engined airliners (i.e. Convair) or six four-engined ones, and this shot was taken in about 1956. That Lufthansa should re-equip with American aircraft in the mid-fifties was due to the dominance of the American industry, helped by the fact that the vanquished Germany was forbidden by the Allies from having its own industry when WW2 ended. Deutsche Lufthansa AG was formed in 1926 becoming known as DLH, and its associated company, the Shanghai-based Eurasia Aviation Corporation, opened up air services in China. Lufthansa expanded under Hitler's National Socialist regime and between 1935 and 1937 total mileage rose by no less than 33 per cent from 10,050,983 to 13,360,422; passengers rose by a surprising 72 per cent from 173,770 to 299,552 and route mileage increased from 27,648 to 39,937. Germany has demonstrated its technical competence over the years, and before WW2 the Junkers company managed to make diesel aero engines work well enough to go into production. This company produced the successful Jumo 205, an extraordinary two-stroke six-cylinder diesel with two crankshafts geared together driven by 12 opposed pistons, giving 600 hp at 2,200 RPM from its 16.63 litres! These Jumo engines powered the twin-engined Junkers Ju 86 airliner used by Lufthansa. Among other machines operated by the airline over the pre-war years were the famous Junkers Ju 52 tri-motor as well as the fast single-engined Junkers Ju 160. Notable centres of pre-war German air routes were, for instance, Berlin, Munich, Hanover, Stuttgart and Hamburg, and DLH cooperated with foreign airlines in its services outside Germany such as KLM, Air France, Ala Littoria and Swissair.

After WW2, Lufthansa no longer operated because of Allied policy on German aviation, but the airline was reformed as Luftag on 6 January 1953 and took its present title of Lufthansa German Airlines in August 1954, starting scheduled Super Connie passenger flights in June 1955 between Hamburg and New York. By early August 1956, the firm's full complement of eight Super-Gs had been delivered (it owned no other version).

B. Robertson

Another shot of a Lufthansa Super-G in the "garage" is shown above. D-ALAK was delivered in April 1955 and served with Lufthansa for three years before being sold to Seaboard & Western in May 1958, but it was bought back by Lufthansa six months later; unfortunately in January 1959 it crashed at Rio de Janeiro on the landing approach and caught fire after a flight from Hamburg. Tip tanks are fitted but

M. J. Hooks

not storm-warning radar; landing lights can be seen behind their transparent covers in the nose and, as usual on radial engines, the cylinders can be seen in the annulus between cowling and propeller spinner. The in-flight shot of Lufthansa's Super Connie D-ALAP shows the machine gaining height after take-off, judging by the machine's attitude, and the afterbodies behind the spinners are clearly visible — the reader may have already noticed that the machine in the previous picture does not have afterbodies! De-icer boots are visible on D-ALAP's wing, tailplane and fin leading edges, and the aircraft appears to be banking to starboard in keeping with the slightly depressed port aileron. With the Turbo Compounds running at climb power, the deep wailing noise from the Super Connie's propellers will be audible on the ground from several miles away, but cabin soundproofing will keep passengers and crew happy. From the purely practical point of

MAP

view, it seemed a pity that Lockheed did not capitalise on the Turbo Compound's modest exhaust note and fit quieter propellers, after all they were available and were fitted to the rival Douglas DC-7, but personally I loved the Super Connie's noise and am very glad that Lockheed saw fit to leave it with its melodious three-blade propellers! In contrast to the climbing Lufthansa machine, the TCA Super-G must be coming in to land because the flaps are much more depressed than they would be for take-off, and the undercarriage would have been retracting or completely up by now if the aircraft were taking off. Note how the flaps effectively increase wing area, and that this Canadian machine has the long nose housing storm-warning radar. Propeller pitch will be set to keep the RPM up giving the impression that the engines are working quite hard even though the machine is descending, as is often the case on aircraft with variable-pitch propellers. Piston engines were quite messy in some respects and it is possible to see the curved exhaust stain on the engine nacelles behind the clean cowlings, and this stain carries on to the flaps which have two thick dark lines on them, one behind each engine. TCA's four Super-Gs were CF-TEU, 'V, 'W and 'X, with CF-TEW being shown here, and these machines were delivered in the order given between April 1956 and December 1957. CF-TEW served with TCA for five years, and all four of TCA's Super-Gs survived to face the breaker's yard. Other Super Connie versions were bought by TCA in the form of five 1049Cs — the first civil versions with Turbo Compound engines (CF-TGA to 'E), three 1049Es (CF-TGF to 'H) and two 1049Hs (CF-TEY and 'Z). In contrast to TCA's machine shown in "dirty" configuration with everything down, Air France's Super-G in the bottom picture (F-BHBB) looks very smooth and the Air France livery on the fuselage goes very well with the curved shape of the eel-like Super Connie.

B. Robertson

MAP

There is no mistaking the fact that TCA's Super-G Connie CF-TEX is taking off in the picture above; the undercarriage is retracting and the folding nosewheel gives the machine a rakish look! The shallow climb angle is noteworthy compared with modern jetliners which seem to go up like lifts, and at full load the Super-G climbed at a leisurely 1,165 ft/min at sea-level and, with one engine out, this fell to

M. J. Hooks

around 650 ft/min. Nervous passengers were assured that airliners could climb away with a failed engine — airliners did, after all, meet engine-out performance criteria — but certain incidents raised doubts about the safety of heavily-loaded machines losing an engine at take-off! I remember over 30 years ago when a large piston airliner, probably a Super Connie or DC-7, lost an engine after take-off and could barely stay in the air even with the good engines running at take-off power. The only way of gaining height was to dump fuel, but the machine was flying over Paris suburbs so it had to stagger along until reaching open ground before fuel dumping; luckily it managed to land safely back at the airport. Those who remember the laboured climb of airliners of that generation, even with all engines running, can but wonder how they could manage on anything less. But fortunately engines were extremely reliable, and in all my hours of watching and flying in airliners during the 'forties and 'fifties I never once saw an engine failure.

CF-TGC was one of TCA's five 1049Cs, a version which was not offered with tip tanks, and is shown lower left flying over an ocean liner. Once upon a time the only way to go, passenger ships were largely displaced by long-haul airliners, the Super Connie included. The busy night shot shows Model 1049H N1880 leased to Lufthansa by Transocean for the German airline's Supercargo service between Frankfurt and New York, which opened on 5 March 1959. The winged emblem on the fin dates back to Deutsche Luftrederei (1919), and note the ceiling lights in the Super Connie's slim fuselage which seems the wrong shape for

M. J. Hooks

cargo other than the self-loading sort! In the bottom picture, a Flying Dutchman is seen landing (look at the flap angle), as Model 1049H *Hermanus Boerhaave* (PH-LKN) kisses the concrete.

MAP

B. Robertson

It was inevitable that the armed forces would find uses for the Connie and Super Connie; indeed the Connie started life as the C-69 military transport but not because it was designed as such. On the contrary the Connie was intended to be an airliner, but the US government banned airliner development when the USA entered WW2 and so the Connie was ordered as a military transport for the USAAF. The machine shown at the top was one of nine military Connie Model 749s delivered to the USAF as C-121As, each with reinforced flooring and a 112 inch by 72 inch cargo door. They could take up to 44 passengers, or 20 stretchers for medical evacuation

M. J. Hooks

MAP

as well as medical staff, but this immaculate machine was used as a Presidential aircraft and was named *Columbine II* (48-610). She appears to have just landed and is turning off the runway; note the short exhaust pipes at the lower part of each engine and the cooling air exits with pronounced cutaways to clear the pipes. The long radar nose and the illusion of a slightly flattened cockpit roof gives the superficial impression that this machine is a Super Constellation. The Super Connie was taken up by the US Navy and USAF in an almost bewildering number of types and subtypes far too numerous to mention here, and the centre picture shows that electronics is not a latter-day invention, as the radar-equipped US Navy WV-2 Warning Star goes about its airborne early-warning work. These machines were assigned the 1049A number by Lockheed, being ordered by the USN in 1950 as PO-2Ws but later becoming WV-2s. They were powered by Turbo Compound engines, their electronic equipment was naturally upgraded during their lives and developments included the WV-2Q for electronics countermeasures and the WV-3 for weather reconnaissance.

A natural role for the Super Connie was that of military transport and the USN ordered the type in August 1950 as the R70-1 transport, to become the R7V-1 as shown in the bottom picture (1049B); it was also powered by Turbo Compound engines as, indeed, were all subsequent piston-engined versions of the Super Connie. The R7V-1 was strengthened, also it had large cargo doors and was even given a sealed floor which could be hosed down.

B. Robertson

This picture shows the prototype Super Constellation (rebuilt from the prototype Connie) fitted experimentally with enormous ventral and dorsal radomes intended for the early warning versions. The upper radome was a sizeable eight feet high, but the ventral one measured 19 by 29 ft and dwarfs the man sitting nearby! "Old 1961", as the much-modified and long-suffering prototype Connie was known, was much used as a development hack by Lockheed and at one time had three different engine types at the same time, these being two Pratt & Whitney R-2800s as port and starboard inners, a Wright R-3350 as the starboard outer and a Turbo Compound in the port outer — it must have sounded most odd! A proposal to offer Connies with Bristol Centaurus engines was not taken up, nor were plans to licence-build Centaurus-powered Connies in Great Britain. How enormous these large aircraft were even in the 'fifties when they were much smaller than large aircraft are now; no wonder they needed 10,000 hp or more to keep them aloft. Incidentally, the standard Connie was always powered by Wright R-3350 engines (or R-3550 Turbo Compounds for the Super Connie), but Lockheed initially offered the Connie with Pratt & Whitney R-2800s, the engines fitted to the Douglas DC-6 and Convair-Liner among other aircraft. But Connie customers, it seems, preferred Wright engines to P & W ones.

MAP

The EC-121T (top) was one of numerous RC and EC-121 variants of the military Super Connie, and AFRES stencilled on the fuselage shows that this one was operated by a USAF Reserve squadron. After WW2, turboprop engines promised a combination of power-to-weight ratio, quiet running, smoothness and reliability simply not possible with piston engines. However, long-haul airliners of the 'fifties managed very well on piston engines, and the only competitive long-haul turboprop airliner, the Bristol Britannia, had engine problems which delayed its entry into service until 1957 — too near the jetliner era for the Britannia to sell well. A few military turboprop Super Connies were built, one version being the R7V-2 shown below with four 5,550 eshp Pratt & Whitney YT-34s; four were produced and the first one flew on 1 September 1954. With over 70 per cent more power than with piston engines the R7V-2 was almost indecently sprightly!

B. Robertson

B. Robertson

The TAP crew lined up beneath their Super Connie Model 1049G follow in the tradition of Portuguese explorer Vasco da Gama, their illustrious compatriot of over four centuries earlier! Propeller spinner afterbodies are fitted and, in addition to the de-icer boots on the wing leading edges, dark-coloured de-icers can also be seen on the propeller blade leading edges up to about two-thirds of the

M. J. Hooks

MAP

radius. Aft of the engine cowlings are cooling air exit flaps with exhaust cutaways, and the nacelles further back are well and truly stained by exhaust deposits. Note also the twin mainwheels with plenty of tread, and the slender three-blade propellers which made their presence so well known as the characteristic Super Connie noise! This noise would be most unenvironmental nowadays, and people living near airports (and not so near) just had to grin and bear it, but the environment was hardly a public issue in those days and presumably only scientists knew what the ozone layer was. No doubt the Super Connie's exhaust helped make this layer what it is today! Relentless commercial pressure from Douglas forced Lockheed to continue developing the Super Constellation, but even the penultimate Model 1049H could not make regular non-stop transatlantic flights in both directions regardless of wind conditions. It fell to Douglas to produce the first truly transatlantic airliner, that is one which could cross the Atlantic non-stop without qualification. This machine, the magnificent DC-7C Seven seas, was developed from the DC-7 series after a request from Pan Am back in 1954, and went into service with this airline on 1 June 1956. Extra fuel was carried in a wing of increased span, and not only was the Seven Seas a long distance airliner *par excellence*, it was fast as well, and although cruising speed is what matters in airliners one of the DC-7C's impressive features was its 400 mph + maximum speed!

Equally impressive were its four Wright Turbo Compound 18EA1 engines giving a massive 3,400 hp each for take-off at 2,900 RPM, driving silent-running Hamilton Standard propellers. Lockheed's reply was a modified Super Constellation, the Model 1649 Starliner shown above and on the previous page, which was designed at TWA's request and which had a completely new wing with greater fuel capacity, among other mods. Span went up to 150 ft from the Super Connie's 123 ft, take-off weight increased from 137,500 lb to as much as 160,000 lb and the engines were mounted just over five feet further outboard. The Starliner had 3,400 hp Wright Turbo Compound R-3350-988TC18EA2s with a maximum propeller speed of 1,015 RPM instead of the Super Connie's 1,269 RPM, a 20 per cent reduction although the tip speed only went down by 11 per cent thanks to a slight increase in diameter; but this cut down the noise. Unfortunately Lockheed's impressive new machine did not enter service until a year after the DC-7C, when TWA introduced the Starliner on 1 June 1957. However this late debut was not the Starliner's only problem because by the mid-fifties airlines were beginning to realise that the future lay with jetliners, and became increasingly reluctant to order new piston-engined machines which would be made prematurely obsolete by commercial jets. So the Starliner failed to sell well and only three airlines bought it, these being TWA which took 29, Lufthansa (4) and Air France (10), the last example of the Starliner being delivered to the French airline in February 1958.

This was the year in which Pan Am took delivery of its first Boeing 707 jetliner (N709PA *Clipper Tradewind*), and by the end of the year it had been joined by five more. This was how things were going to be, and the fascinating battle for supremacy and the relentless development of piston-engined long-haulers was now at an end. Douglas stayed in the airliner race with its DC-8 jetliner, but its new sparring partner was Boeing because Lockheed had no jetliner to offer and dropped out of the running until its Tristar came along some years later. So the beautiful and dignified Starliner was the swansong of the colourful piston airliner era which had started over a quarter of a century before, but those of us who were privileged to witness the Lockheed-Douglas battle will remember it with awe, and not without a touch of regret that it came to an end.

Technical Data

Constellation Models 049 to 749A

Type: Four-engined airliner.

Wings: Cantilever low-wing monoplane type. All-metal aluminium-alloy two-spar structure with flush-rivetted stressed-skin. Hydraulically-boosted ailerons. Lockheed-Fowler flaps housed in trailing edge recesses, increasing wing-area when deployed.

Tail unit: Cantilever monoplane type with three fins and rudders. Fixed surfaces of all-metal stressed-skin structure. Control surfaces have fabric-covered metal frames. Hydraulic boost.

Fuselage: All-metal semi-monocoque structure of circular cross-section.

Accommodation: Various arrangements to suit different operators. Typical capacity 47 passengers and 7 crew (049), 48 day/24 night passengers and 7 crew (649, 749, 749A), 57 passengers (649A). Soundproofed pressurised cabin with 8,000 ft (2,438 m) effective cabin altitude at 20,000 ft (6,096 m) aircraft altitude. Thermostatically-controlled heating/cooling. Freight space.

Engines: Four Wright R-3350 Cyclone 18 series supercharged 18-cylinder aircooled radial engines driving three-blade constant-speed propellers through 0.4375:1 reduction gear. Reversible-pitch propellers on 649 onwards. Bore and stroke are 155.6 mm and 160.2 mm respectively (54.56 litres). Take-off power 2,200 hp (049) and 2,500 hp (649, 649A, 749, 749A) at 2,800 RPM from, respectively, C18BA series engines and 749C18BD1 engines.

Landing gear: Retractable tricycle type with twin-wheels on nose and main undercarriage legs. Single compression leg on nose and mainwheels, latter fitted with hydraulic braking system. Steerable nosewheel. Mainwheels retract forwards into inner engine nacelles, and nosewheel retracts backwards into fuselage.

Dimensions: Span 123 ft (37.49 m), length 95 ft 3 in (29.03 m), wing area 1,650 ft^2 (153.29 m^2). Cabin length x max width x max height 64 ft 9 in x 10 ft 8.6 in x 6 ft 6 in (19.73 m x 3.27 m x 1.98 m). Mainwheel track 28 ft (8.53 m). Freight volume 445 ft^3 (12.60 m^3).

Weights & Loadings: (049) Maximum 86,250 lb (39,123 kg), max wing loading 52.27 lb/ft^2 (255.21 kg/m^2), max power loading 9.8 lb/hp (4.45 kg/hp).

(649) Maximum 94,000 lb (42,638 kg), max wing loading 56.97 lb/ft^2 (278.16 kg/m^2), max power loading 9.4 lb/hp (4.26 kg/hp).

Performance: (049) Cruising speed 313 mph (504 kph) at 20,000 ft (6,096 m), landing speed 79 mph (127 kph), rate of climb at sea-level 1,620 ft/min (8.23 m/s), still-air range with max fuel/no reserves 3,991 miles (6,422 km).

(649) Cruising speed 327 mph (526 kph) at 20,000 ft (6,096 m), landing speed 87 mph (140 kph), rate of climb at sea-level 1,420 ft/min (7.21 m/s).

Super Constellation Models 1049 to 1049H

Type: See 049 to 749A[1].

Wings: See 049 to 749A[1]. However, integrally-stiffened skin panels on later aircraft.

Tail Unit: See 049 to 749A[1]. However, elevators have all-metal stressed-skin structure.

Fuselage: See 049 to 749A[1].

Accommodation: Passenger-carrying versions have various arrangements to suit different operators. Typical capacity 47 to 82 passengers and 11 crew. Soundproofed pressurised cabin with 8,000 ft (2,438 m) effective cabin altitude at 22,800 ft (6,949 m) aircraft altitude. Thermostatically-controlled heating/cooling. Freight space.

Engines:	(1049) Four Wright R-3350-956C18CA1 Cyclone 18 supercharged 18-cylinder aircooled radial engines driving three-blade constant-speed propellers through 0.4375:1 reduction gear. Reversible-pitch propellers. Bore and stroke as for 049 to 749A. Take-off power 2,700 hp at 2,900 RPM. Later on 975C18CB1 engines used giving 2,800 hp at 2,900 RPM.

(1049A to E, G, H) Four Wright R-3350 Turbo Compound supercharged 18-cylinder aircooled radial engines driving three-blade constant-speed reversible-pitch propellers through 0.4375:1 reduction gear. Three exhaust-driven aircooled power-recovery turbines equally disposed about rear of engine, each supplied by six cylinders and coupled to crankshaft through gearing and hydraulic drive. Bore and stroke as 049 to 749A. Take-off power 3,250 hp (1049C, D, E and G) and 3,400 hp (later 1049G and H) at 2,900 RPM from, respectively, TC18DA series engines and EA series engines.

Landing Gear: See 049 to 749A[1].

Dimensions: Span 123 ft (37.49 m)[2], length 116 ft 2 in (35.41 m), wing area 1,650 ft^2 (153.29 m^2). Cabin length x max width x max height 83 ft 2 in x 10 ft 8.6 in x 6 ft 6 in (25.35 m x 3.27 m x 1.98 m). Mainwheel track 28 ft (8.53 m). Freight volume (passenger-carrying versions) 693 ft^3 (19.62 m^3).

Weights & Loadings: (1049) Maximum 120,000 lb (54,432 kg), max wing loading 72.73 lb/ft^2 (355.10 kg/m^2), max power loading (2,800 hp engines) 10.71 lb/hp (4.68 kg/hp).

(Standard 1049C) Maximum 133,000 lb (60,329 kg), max wing loading 80.61 lb/ft^2 (393.58 kg/m^2), max power loading 10.23 lb/hp (4.64 kg/hp).

Performance: (1049 with 2,800 hp engines) Cruising speed 301 mph (479 kph) at 20,000 ft (6,096 m), landing speed 95 mph (153 kph).

(Standard 1049C) Cruising speed 314 mph (505 kph) at 20,000 ft (6,096 m), landing speed 97 mph (156 kph), rate of climb at sea-level 1,125 ft/min (5.72 m/s), still-air range with max fuel/no reserves 4,754 miles (7,649 km).

Starliner Model 1649A

Type: See 049 to 749A[1].

Wings: Cantilever low-wing type. Integrally-stiffened all-metal structure. Hydraulically-boosted ailerons. Lockheed-Fowler flaps housed in trailing-edge recesses, increasing wing area when deployed.

Tail Unit: See 1049 to 1049H[3].

Fuselage: See 049 to 749A[1].

Accommodation: Various arrangements used, examples being 68 passengers and 11 crew (TWA), 62 passengers and 11 crew (Air France, Lufthansa). See 1049 to 1049H for remaining description[3].

Engines: Four Wright R-3350-988TC18EA2 Turbo Compound supercharged 18-cylinder aircooled radial engines driving three-blade constant-speed reversible-pitch propellers through 0.35:1 reduction gear. Power-recovery turbine description as for 1049 to 1049H[3]. Bore and stroke as for 049 to 749A. Take-off power 3,400 hp at 2,900 RPM.

Landing Gear: See 049 to 749A[1].

Dimensions: Span 150 ft (45.72 m), length 116 ft 2 in (35.41 m), wing area 1,850 ft^2 (171.87 m^2). Cabin internal dimensions as for 1049 to 1049H. Mainwheel track 38 ft 4.8 in (11.70 m). Freight volume 593 ft^3 (16.79 m^3).

Weights & Loadings: Maximum 156,000 lb or 160,000 lb (70,762 kg or 72,576 kg) depending on propellers, max wing loading (156,000 lb weight) 84.32 lb/ft^2 (411.69 kg/m^2), max power loading (156,000 lb weight) 11.47 lb/hp (5.20 kg/hp).

Performance: Cruising speed 342 mph (550 kph) at 20,000 ft (6,096 m), landing speed 101 mph (163 kph), rate of climb at sea-level 1,080 ft/min (5.49 m/s), still-air range with max fuel/no reserves 6,100 miles (10,103 km).

Notes:
[1] 049 to 749A description applies, other features may differ.

[2] 123 ft 5 in with tip tanks (1049G and H).

[3] 1049 to 1049H description applies, other features may differ.